Florida
Plants and Animals

Bob Knotts

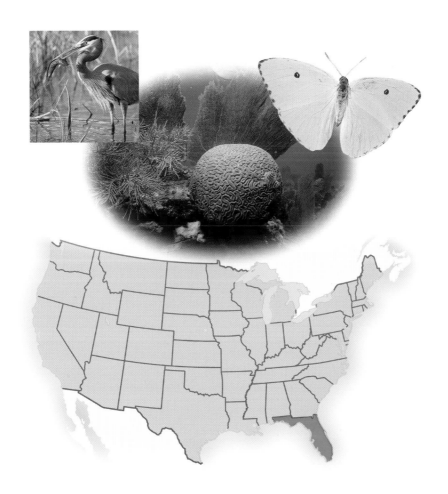

Heinemann Library
Chicago, Illinois

© 2003 Heinemann Library
a division of Reed Elsevier Inc.
Chicago, Illinois

Customer Service 888-454-2279

Visit our website at www.heinemannlibrary.com

Designed by Heinemann Library
Page layout by Wilkinson Design
Printed and bound in the United States by Lake
Book Manufacturing, Inc.

07 06 05 04 03
10 9 8 7 6 5 4 3 2 1

**Library of Congress
Cataloging-in-Publication Data**

Knotts, Bob.
  Florida plants and animals / Bob Knotts.
    v. cm. -- (State studies)
Includes bibliographical references (p.47).
Contents: Florida's plants and animals -- Florida's
habitats --
Endangered plants and animals -- Extinct animals.
  ISBN 1-40340-350-3 (hc) -- ISBN 1-40340-566-2
(pb)
  1. Natural history--Florida--Juvenile literature. [1.
Zoology--Florida. 2. Botany--Florida.] I. Title. II.
State studies
(Heinemann Library (Firm))
  QH105.F6 K66 2002
  578'.09759--dc21

                                          2002005711

**Some words are shown
in bold, like this. You can
find out what they mean
by looking in the glossary.**

Acknowledgments
The author and publishers are grateful to the
following for permission to reproduce copyright
material:

Cover photographs by (top, L-R) L. Lipsky/
Bruce Coleman, Inc., Joel Arlington/Visuals
Unlimited, Stephen J. Nesius/Heinemann Library,
Joe McDonald/Visuals Unlimited, (main) Tony
Arruza; title page (L-R) J.H. Robinson/Photo
Researchers, Len Kaufman, Bob Gossington/Bruce
Coleman, Inc.; contents page (main) David R.
Frazier Photolibrary, (inset) John Shaw/Bruce
Coleman, Inc.; p. 4 Farrell Grehan/Photo
Researchers; pp. 5, 11, 27, 33, 45 maps.com/
Heinemann Library; p. 6 Tony Arruza; p. 7 Kevin
Fleming/Corbis; p. 8 David R. Frazier Photolibrary;
p. 9 L. Lipsky/Bruce Coleman, Inc.; p. 10 Christine
Taylor/Florida International University; p. 12T J.H.
Robinson/Photo Researchers; p. 12B Carrie
Gowran; pp. 13, 38 Stephen J. Nesius/Heinemann
Library; p. 14T M.H. Sharp/Photo Researchers;
p. 14B Rick Poley; p. 15 Joel Arlington/Visuals
Unlimited; p. 16 Lee Foster/Bruce Coleman, Inc.;
p. 17 Janet Hinchee/USDA Forest Service;
p. 18 David Sieren/Visuals Unlimited; p. 19 Eric
and David Hosking/Corbis; p. 20 John Shaw/Bruce
Coleman, Inc.; p. 21 David and Hayes Norris/
Photo Researchers; p. 22 Wendell Metzen/Bruce
Coleman, Inc.; p. 23 Len Kaufman; p. 24 Hal
Beral/Visuals Unlimited; p. 25 Barry Parker/Bruce
Coleman, Inc.; p. 26 Brandon D. Cole/Visuals
Unlimited; p. 28T Marc Epstein/Visuals Unlimited;
p. 28B Bob Gossington/Bruce Coleman, Inc.;
pp. 29, 42 Jeff Greenberg/Photo Researchers;
p. 30T David S. Addison/Visuals Unlimited; p. 30B
David Sieren/Visuals Unlimited; p. 32T Joe
McDonald/Visuals Unlimited; p. 32B Tom and Pat
Leeson/Photo Researchers; pp. 34, 37 Used with
permission of University Press of Florida/mural by
Marisa Renz, photo by Mark Renz; p. 35 Mary
Evans Picture Library; p. 36 Historical Picture
Archive/Corbis; p. 39 Laura Riley/Bruce Coleman,
Inc.; p. 40 Stephen Kline/Bruce Coleman, Inc.;
p. 41 J.P. Jackson/Photo Researchers; p. 43 S.
Maslowski/Visuals Unlimited; p. 44 James
Beveridge/Visuals Unlimited

Photo research by Julie Laffin.

Special thanks to Gail H. Compton (Naturalist,
St. Augustine Lighthouse & Museum) and Frank
C. Watts (Soil Survey Project Leader, Natural
Resources Conservation Service, U.S. Department
of Agriculture) for their comments in the
preparation of this book.

Every effort has been made to contact copyright
holders of any material reproduced in this book.
Any omissions will be rectified in subsequent
printings if notice is given to the publisher.

Contents

Florida's Plants and Animals

Imagine a place with beautiful beaches and tall, swaying palm trees. It is a warm, sunny place most of the year, but it also gets a lot of rain. This place is the state of Florida, which is called the Sunshine State. It sits in the far southeastern corner of our country.

FLORIDA'S PLANT AND ANIMAL LIFE

Florida's sunshine and rain help many plants grow. This land, with lots of plants, is also a good place for insects and small animals to live. They can find food easily among all those plants. The many smaller **herbivores** then make it easy for bigger animals to

Gulls are just one of many forms of life found in the Sunshine State. They usually make their home near the ocean. Gulls eat fish and other water animals, insects, rotten meat, and the eggs and young of other birds.

Florida and the Southeastern United States

The states of Georgia and Alabama are north and west of Florida. Since the state is surrounded by ocean waters on three sides, it is a peninsula.

find food, too. This is because the larger animals are often **carnivores** that eat the smaller animals.

This same process happens beneath the lakes and ocean waters that are so common all around Florida. Many plants grow underwater. These water plants make good food for a lot of small creatures, including fish. Then many kinds of bigger fish and other water wildlife eat the smaller animals.

On land and under the water, Florida is full of living creatures, large and small. It is a special place to study plants and animals. There is nowhere else on Earth just like it.

PLANTS

A variety of plants live in Florida because of the sunny weather. Plants use sunshine to make their food. Florida has more kinds of trees than any other state in this country. More than 300 different trees grow in Florida.

Florida has smaller plants, too. At least 6,000 different plants grow in this state. Some are leafy green shrubs. Others are tall and thin plants. There are also vines and wildflowers of many different types.

ANIMALS

The warm weather in Florida makes it easy for many types of animals to live in the state. There are animals of all kinds in the water, under the water, and on the land. Alligators live in this state, and there are even some crocodiles at the southern tip of Florida. Sharks live in the ocean around Florida. Many kinds of fish live in Florida's lakes and streams. Still other kinds of fish live in the sea.

Not all trees in Florida are palm trees. The state also has pine forests. These pine trees are located on Santa Rosa Island, but there are many more throughout the state.

Large birds live in Florida, such as bald eagles and flamingos. Both poisonous and nonpoisonous snakes and spiders crawl or slither around the earth.

*Flamingos have long, stiltlike legs and curved bills and necks. They grow to a height of three to five feet tall, and their feathers are pink or red in color. Flamingos eat **shellfish** and algae and spend their entire lives near water.*

EVERGLADES

One of Florida's special **habitats** is called the Everglades. The Everglades is an area that covers a large part of southern Florida. There is no other spot in the world like the Everglades.

Everglades Bugs

If you visit the Everglades, you probably should go sometime between late fall and early spring. In warmer weather, this swampy place has many insects that might make your visit no fun at all.

Mosquitoes by the millions live in the Everglades during the summer months. So do flies of different kinds and other pesky bugs. They like the heat and rain that summer brings to southern Florida.

The cooler, drier weather of late fall and early spring helps keep the number of Everglades insects much lower.

People sometimes fish and hunt in the Florida Everglades. Mostly, however, this special place is home to many species of plants and animals.

The Everglades look like a giant swamp, with shallow water covering the land. Native Americans called this place "grassy waters" because tall, grasslike plants grow throughout the Everglades. Today, many people think of the Everglades as a river of grass, because it really is a very wide, slow-moving river. The water is freshwater, which means it has little or no salt in it. Freshwater is different from saltwater, which is found in the ocean. The animals and plants that live in freshwater **habitats** are different from those living in saltwater habitats. Much of the Everglades water is only about six inches deep, flowing from north to south toward the ocean.

Habitat Variety

Florida has many other important habitats. Some of these places are narrow, sandy areas such as the state's many beaches. Other places are under the water, including

Florida's amazing coral **reefs.** Coral reefs look like beautifully colored rocks under the ocean, but they are really made of tiny living sea creatures called **coral polyps.** Still other habitats are covered with tall pine forests.

This book explores the many different habitats of Florida. In your reading, you will learn more about the plants and animals that live in these habitats. You will even learn something about the animals and plants that lived in this land thousands of years ago.

A trip around Florida is an adventure filled with strange creatures, both large and small. If you are ready to discover them now, the wonderful plants and animals of Florida are waiting for you!

What Is a Habitat?

One important reason Florida has so many different plants and animals is that the state has many different areas for them to live in. These areas are called habitats. A habitat is a place that provides exactly the things certain kinds of animals or plants need to live there.

Colorful schools of fish live around the coral reefs off Florida's shores. This unusual ocean habitat is as much a part of Florida as are the beaches and swamps.

Florida Habitats

There are many different **ecosystems** in Florida. Each has its own special types of plants and animals. The climate and **natural resources** of each **habitat** help to determine what kinds of plants and animals live there.

EVERGLADES

The swampy area called the Everglades is full of water, and all of the wildlife there depends on the water in one way or another to survive. Some plants and animals live in the water. Others find their food in the water. For example, a heron, a large bird, doesn't live in the water, but it wouldn't be able to live without the food it finds there.

*Blue-green algae are a part of the **food chain** in the Everglades. They help clean the air and water and serve as food for fish and other water animals.*

Florida Habitats

0 _____ 100 mi.

0 _____ 100 km

N
W   E
S

GULF OF MEXICO

ATLANTIC OCEAN

Pensacola
Tallahassee
Suwannee River
St. Mary's River
Jacksonville
St. Johns River
St. Augustine
Apalachicola
Lake George
Daytona Beach
Orlando
Tampa
St. Petersburg
Sarasota
Lake Okeechobee
West Palm Beach
Fort Lauderdale
Miami
Key Largo
Key West

Beaches and Sand Dunes
Coral Reefs
Mangroves
Everglades
Big Scrub
Wetlands
Pine Forests
Mixed Habitats

The Everglades is full of life. Water plants called algae are a type of life you will notice in the Everglades. Algae look a bit like stringy mosses. Algae are found as part of periphyton (pear-if-eye-ton), which are communities of **microorganisms** that cover the surface of much of the water in the Everglades.

In the Everglades, tiny creatures like mosquito **larvae, tadpoles,** salamanders, and small fish eat algae. Then, larger creatures such as bigger fish and frogs eat the smaller creatures. Still larger animals then eat the frogs

*Florida's Everglades make up a **unique environment,** but Florida has many different habitats, each with its own community of plants and animals.*

11

Great blue herons feed on fish in both fresh and salt water. They usually hunt alone, but nest in flocks.

and fish. These larger animals include big fish, turtles, large birds, such as the great blue heron, and mammals. Finally, the biggest animals, such as the alligators, eat these large birds, fish, turtles, and other creatures.

That process is called the **food chain.** It shows how important all the plants and animals in a **habitat** are. Each needs the others to survive. If some of the smallest forms of life died, the largest animals would probably die, too. In the Everglades, for instance, the large alligators would eventually suffer if the tiny algae suffered. Because animals often eat more than one thing, they may appear in more than one food chain. Within an **ecosystem,** these different food chains are joined to form a food web. An ecosystem includes all of the animals and plants that make up a particular community living in a certain **environment.**

*This Florida food chain is part of a larger food web. In this food chain, the plant is a **producer,** and the deer and alligator are **consumers.** The plant produces energy, and is then consumed by the deer. The deer is then consumed by the alligator. The energy produced by the plant passes on through the food chain.*

THE ALLIGATOR

Alligators have **evolved** perfectly to live in Florida. Since they are very dark in color, they often look like floating logs. This lets them sneak up on other animals. That dark color would not work in a place where the water was clear, as other animals could see the alligator approaching.

Another **adaptation** found on an alligator is its third eyelid. Their top and bottom lids close like ours, but the third lid is clear and protects their eyes underwater. Alligators can swim underwater and see animals they want to eat.

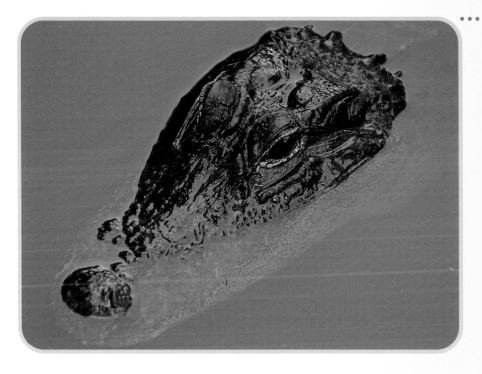

Alligators' eyes stick up above their skulls so that they can see above the water while their bodies are underwater. Visitors to the Everglades often see alligators, but these creatures usually prefer to stay away from humans.

Alligators are an important part of the Everglades. For example, they make "gator holes," which help smaller animals survive through the dry, winter months when there is less water available. To make these gator holes, alligators clear plants and soil from large holes in the ground just below the water by digging with their wide noses and strong tails. Then, they live in these holes when winter comes.

Florida river otters depend on gator holes to help them survive dry weather. Otters are excellent swimmers, but don't move nearly as well on land. They have special muscles that allow them to close their ears and nostrils to keep water out.

When the Everglades become drier in winter, gator holes turn into small ponds. They are deeper than the land around them and hold water like a bathtub. Other animals, like frogs and fish, live in the gator holes. Deer, birds, panthers, river otters, and mink come to the holes to drink and eat when the alligators are not home. In this way, the alligators help a lot of other animals live through the dry season. This is another example of how all the life in this habitat is connected.

Florida's Crocodiles

Crocodiles look a lot like alligators, but their backs are a lighter color and their noses are thinner. Florida is the only state in this country where crocodiles live, though they are rare even there. Scientists think that between 600 and 1,000 crocodiles live around the  mangroves at the southern end of the state, and in some of the northern Keys.

The habitat of the crocodile includes a small part of the Everglades. That makes the Everglades the only area on Earth where both crocodiles and alligators live.

Everglades Plants

The Everglades also support **unique** trees and plants. Just below the surface of the Everglades lies a layer of rock called limestone. This rock sticks out of the water in some places, and soil collects there. This means that pine trees and palm trees are able to grow.

The saw palmetto is one of the most common plants in Florida. It grows wild in the woods, but is also seen around many homes. Its fan-shaped leaves can be used to make baskets.

Other common Everglades plants include the saw palmetto, which looks like a very small palm tree, and Spanish moss, which often hangs in trees. Spanish moss is an **epiphyte.** Trees such as cocoplum and sweetbay magnolia are also often seen in wet areas. What these plants all have in common is their need for the plentiful water that is found in the Everglades **habitat.**

Mangrove Habitats

Mangroves are another common habitat in Florida. These areas are named for the red, black, and white mangrove trees that grow where the water and land meet.

Mangrove trees look like bunches of thick sticks all tangled together in the water. They live in saltwater, including the sea at the edge of the Everglades. Mangroves also grow throughout the chain of islands called the Florida Keys, at the southern end of the state, and all the way up to Daytona Beach in Volusia County. Mangrove trees are able to live in saltwater because they get rid of the salt through their leaves. This lets them live in an **environment** where most other trees cannot grow.

*Mangrove trees like these have **adapted** to their environment near Florida's coasts. They have long, spreading roots that find and hold onto the dirt and sand of the area, which helps build up shorelines.*

The tangled roots of red mangroves are very important for many animals. For example, fish like the mangrove snapper raise their young among these roots, as do frogs. The roots offer a great place to hide from larger **predators** who might eat the young animals.

Mangroves and other trees in this part of Florida make good homes for many large birds, which build their nests in these trees. Young crabs, shrimp, and other sea creatures also live among Florida's many mangroves.

THE BIG SCRUB

The odd-sounding Big Scrub is a ridge of sandy land in the center of Florida. Like the Everglades, the Big Scrub is **unique.** Thousands of years ago, Florida was mostly covered with water. During that time, the Big Scrub was on the coastline. In other words, it was once a beach.

Much later, more of the land became dry as some of the ocean water evaporated and the shoreline moved further out. The Big Scrub was then in the middle of Florida. Scientists think that the plants and animals in this **habitat** may be the oldest in the state.

Today, the Big Scrub is very dry nearly all year long. In this way, it is almost a desert. Only plants and animals that need little water can live there.

*The Big Scrub is one of the most interesting and **endangered** natural plant communities in the United States. It is also home to many plants and animals that are not found anywhere else in the world.*

The sandy soil and dry climate are perfect for many trees. For example, this is the only area in the world where sand pines grow wild. Other trees in the Big Scrub that enjoy the sandy soil are the bluejack oak, the myrtle oak, and the chapman oak. There are some small flowering trees that only grow in the Big Scrub, including the pygmy fringe tree.

Many animals live in the Big Scrub. Some are common animals, such as raccoons, squirrels, black bears, and wild turkeys. They can find food easily in this sandy place. Other animals look more like those you would expect to find in a desert, like the scrub lizard, gopher tortoise, diamondback rattlesnake, and indigo snake.

As in the Everglades, all these animals and plants are important parts of this dry Florida habitat. Each one helps to keep nature in balance by feeding on other animals or plants that live there. In turn, many of these animals and plants become food for other creatures. It is all part of nature's cycle of life.

PINE FORESTS

Many people think of palm trees when they think about Florida. But did you know that more than half of this state has pine trees on the land? There are many types of pine trees. In some parts of Florida, the pine trees grow close together to make thick forests. For example, northern Florida has many pine forests.

One of Florida's important pines is the slash pine. This tree can be found in almost all areas of Florida. It is used as

Longleaf pines have thick, fire-resistant bark, long needles, and large cones. Their needles can grow up to eighteen inches long. The wood of longleaf pines is strong and heavy, and is often used to build houses.

The slash pine's needles and cones are smaller than those of the longleaf pine. The slash pine grows rapidly and lives approximately 200 years.

pulp wood to make paper, and is also one of this country's main sources of **turpentine.** Another of Florida's pines is the longleaf pine. These trees can grow for 500 years, and their trunks can grow up to two feet thick.

Florida's pine forests are important because other, smaller plants live in these same areas. The pine trees have long, thin needles that let sunshine reach the ground. Therefore, the ground under the pine branches is perfect for plants such as wiregrass, rabbit tobacco, beggarweed, fetterbush, and saw palmetto.

All these trees and plants support a population of insects that are specially suited to this environment. Many insects, such as grubs and beetles, live in trees that have fallen to the ground. The fact that there are so

many insects helps to bring birds to the forests, too. Birds such as warblers and woodpeckers eat these insects. Great horned owls are right at home in these pine forests, where they find plenty of their favorite foods, such as smaller birds, snakes, and rodents.

You can also find some of Florida's larger animals in a pine forest. They are at the top of the **food chain,** eating smaller animals and plants. If you are lucky, you might spot a black bear catching fish in a river, or a white-tailed deer eating berries.

Great horned owls hunt for food at night in Florida's pine forests. The great horned owl is the only large owl of North America that has tufts of feathers on its head. This owl can grow up to two feet tall.

Reptiles and **amphibians** are other kinds of animals that live in this state's pine forest **habitat.** The poisonous diamondback rattlesnake and the box turtle are among nearly 30 kinds of reptiles and amphibians that live in the pine forests, feeding on smaller animals and plants.

BEACHES AND SAND DUNES

If you have been to an ocean beach before, you may have noticed shells that washed in with the waves. However, you probably didn't know that many plants and animals live on beaches and in sand dunes. This is especially true in Florida, where the warm weather helps many things grow. For instance, plants called sea oats grow in the Florida dunes. Their long roots spread out and help keep the sand from being blown or washed away.

Many kinds of animals like to live on the beach, too. Insects like the tiny beach flea, or sandhopper, live in seaweed that has been washed ashore by the tide. Larger animals then feed on these insects. Many animals such as crabs, which are common on Florida beaches, live on or near the beach because they need seawater to live. Crabs have hard shells and ten legs, including strong claws in front. They crawl along the sand looking for food. Florida's crabs include the horseshoe crab and the hermit crab.

The beach is also a good habitat for many birds. Gulls of different kinds fly over beaches and sand dunes looking for food. Gulls like to eat fish most of all, but they sometimes also eat leftovers, including foods like French fries, left by people who visit the beach. However, most human foods are not good for the gulls.

One of the most interesting beach animals is the sea turtle. The largest is the Atlantic leatherback, which can be 6

*Florida's beaches may look almost empty of life, but a closer look shows many **species** of plants and animals. Sea oats look like tall grass, and grow up to three or four feet tall.*

Very few people are lucky enough to watch sea turtles on the beach in Florida. This female Atlantic loggerhead turtle is nesting and laying her eggs.

feet long and weigh 1,300 pounds. Other common sea turtles include the Atlantic loggerhead and the Atlantic green turtle. These large turtles live in the ocean for most of the year, but the female turtles must return to the beach to lay their eggs. They crawl slowly into the sand dunes at night to do this. When the baby turtles, called hatchlings, come out of their eggshells, they crawl by moonlight to the ocean.

CORAL REEFS

In the **continental United States,** only Florida has coral **reefs.** Florida's coral reefs are a **unique** habitat entirely underwater. Some of Florida's reefs are found near the southeast coast, close to Fort Lauderdale and Miami. Most of the reefs are near the Florida Keys. There are about 6,000 coral reefs in this area, and each is a whole different world of unusual underwater animals.

When you first see a coral reef, it looks like a collection of differently shaped rocks in many pretty colors. Yet a coral reef is not made of rocks. Every reef is made from millions of tiny sea animals called **coral polyps.** These animals **evolved** to be specially suited to live in warm, clear seas like those around Florida's southern coast.

Coral polyps are very tiny animals that are shaped like small tubes. They have mouths with little **tentacles** that catch tiny sea animals, including very small shrimp and fish. They sting their prey with these tentacles, then eat it. Coral polyps make their own shells by giving off a substance from their bodies. The substance is called calcium carbonate. Calcium carbonate is the same thing that helps make our teeth and bones.

As many polyps build their hard shells, these shells begin to grow together. This takes a long time since some kinds of coral only grow one-half inch every year. But as the years pass, the shells of the coral polyps start to make different shapes. It can take thousands of years to make a coral reef as large as those in Florida. This is one reason why coral reefs are so special.

Some kinds of coral, including finger coral, brain coral, rose coral, and staghorn coral, look very hard. Other types, such as sea plumes, sea whips, and sea fans, look like plants and sway with the ocean currents underwater. All coral is brightly colored and unusually shaped.

Many types of coral make up the reefs along Florida's coast. Divers enjoy viewing the different colors and shapes, but they must be careful not to damage the **fragile** coral reef. These gorgonian (left) and brain (right) corals are found in John Pennekamp Coral Reef State Park near Key Largo, Florida.

Coral **reefs** are a very important home for different types of sea creatures. Fish, including angelfish, zebra fish, and parrot fish, live near coral reefs. They find their food there and use the coral to hide from **predators.**

Other animals live around Florida's coral reefs because they can easily find food there. These animals include the barracuda, a very fast, thin fish with big teeth, as well as crabs, lobsters, and snails.

The shark is one of the best-known animals around Florida's coral reefs, and is also the largest and strongest. There are many kinds of sharks that live near

The Unusual Parrot Fish

The parrot fish looks just like its name—like a parrot. There are about fourteen kinds of parrot fish off the coast of Florida. These brightly colored fish are often blue and green. Their mouths are usually black, and look like parrots' beaks. The mouths of parrot fish are strong and hard like birds' beaks. These fish feed on coral, algae, and sea worms. If you swim underwater near parrot fish, you can actually hear the sound they make while chewing.

By crunching up coral with their mouths, parrot fish make sand. Some of this sand later gets washed onto the shore, and this helps build Florida's beaches.

*This pregnant nurse shark was photographed in Florida waters. Unlike most types of sharks, the nurse shark can pump water over its **gills**. Therefore, it does not have to swim constantly in order to breathe and can lie motionless on the sea floor, where it finds food.*

Florida's coral reefs. These sharks look very similar when you first see them, but there are differences.

The Caribbean shark is a little over five feet long and will occasionally attack people. Though the nurse shark can grow larger than this, it is usually gentle and rarely bothers people. Other sharks found around Florida's coral reefs include the tiger shark, bull shark, spinner shark, and blacktip shark.

Most sharks don't bother people unless people bother them first, but they sometimes bite human beings by mistake. The water may be so cloudy that a shark might think an arm or leg is a fish. Or, a shark might be attracted to a person who is wearing something shiny, which can look like fish scales to a shark. Sharks would much rather eat fish than people, though. In the United States, most shark attacks happen in Florida. Since 1990, there have been more than 385 shark attacks on people in the United States. Of these, 250 took place around Florida.

Coral is clearly important to many different animals. Unfortunately, these reefs can be easily harmed. Spilled oil from boat motors can hurt coral, and anchors from boats, scuba divers, and snorkelers can scrape or break the coral reef. No person should ever touch living coral. If they do, they can cause an infection in the coral. This is the same kind of infection you might get if you cut your finger and didn't wash it. In coral, just as in people, the infection can spread. In time, that can kill the coral.

If left alone by people, coral reefs will keep growing. They will bring all kinds of fascinating sea creatures to the Florida coast for many years to come.

Manatees are also called sea cows. They eat more than 100 pounds of plants a day, grow to 13 feet long, and weigh up to 3,500 pounds! In the winter, manatees often swim into canals near power plants, where warm water from these facilities is released.

MANATEES

Florida has many other **unique habitats.** Florida's **manatees,** which are large, gentle, water mammals that move slowly, need the state's warm spring waters in order to live. Springs are pools of water that bubble up from underground. Florida has more than 600 springs, more than anywhere else on Earth. These springs stay about 72°F (22°C) all year long. That makes them good places for manatees to live during the colder winter months. Manatees feed on plants that live in the water. If the water temperature drops to around 60°F, however, manatees will stop eating. They can also get sick from the cold water.

WETLANDS

The wetland habitat is very important to Florida. The

Florida Wetlands

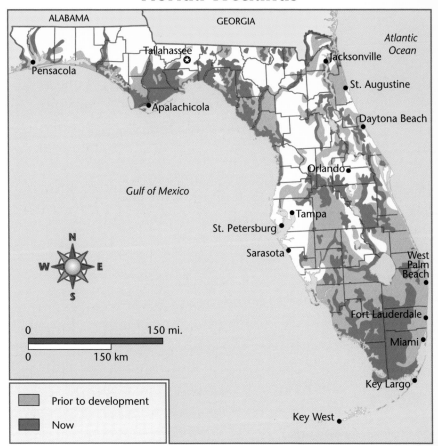

This map shows where Florida's wetlands have been lost to development over the years. Many people are now trying to save Florida's remaining wetlands.

state's wetlands are places where the land is covered with water for at least one month every year. These are damp places, but different from the swampy Everglades, which offer some water all year long.

Plants that need a lot of water to grow can live easily in wetlands. For example, cattails and water lilies are common in wetland areas. Cypress and mangrove trees also grow here. Marsh grasses poke through the soil, adding a little green color to the watery landscape. Visitors to the wetlands may also find ferns, which are small green plants that have leaves but no flowers.

Animals of the wetlands include mosquitoes, dragonflies, **tadpoles,** water bugs, and snails. Large birds like the great egret gather there to eat these small wetland creatures. Raccoons, water snakes, and turtles also enjoy the wide range of things the wetlands offer to eat.

The wax myrtle is also called the bayberry tree. It grows up to 35 feet tall, has small green flowers, and has leaves that grow to be 2 to 4 inches long.

OTHER COMMON PLANTS AND ANIMALS

Of course, many other plants and animals live in Florida. The wax myrtle is a common Florida plant. This grows as a bushy tree or shrub and attracts birds and other small animals. It is dark green with gray berries. Bluebirds, swallows, bobwhites, chickadees, and many other birds love to eat the berries of the wax myrtle tree.

One of Florida's most common animals is the gray and white mockingbird, Florida's state bird. The mockingbird sings many songs as it imitates, or mocks, the songs of other birds. There are also opossums, rats, and bats, as well as beautiful butterflies, such as the yellow sulfur butterfly and the orange and black monarch. Florida is so warm for most of the year that insects live in all parts of the state. These include cockroaches, termites, and spiders.

Remember that no wild animal wants to be near humans. They only bite or sting when they are bothered or frightened. Each year, human beings move into more natural places in Florida. If future generations are to see Florida's wonderful wildlife, we must take care in the present to maintain their natural habitats.

Florida's butterflies, like the sulfur butterfly at left, are able to live year-round in the state's warm weather. Sulfur butterflies range from light yellow to orange in color and are named after the powdery yellow mineral called sulfur.

Endangered Plants and Animals

Usually, animals and plants are **endangered** because of things people do. We sometimes hunt animals so much that we kill almost all of them. Or, we may build homes and businesses where this wildlife has always lived. By doing this, we accidentally kill some of them and force others to move to **habitats** they are not used to. Being in an unfamiliar habitat makes it harder for the animals to find their food and raise their young. Therefore, more of them are likely to die.

Florida has more than 500 endangered plants and animals. Some are large, powerful animals that need a lot of land to live. Others are small plants or animals that only live in a few special places.

People continue to build new housing in wetland areas of Florida. This puts animals and plants in danger by destroying their habitats.

This hand fern is lucky to have a place to live. Many ferns are losing their habitats in Florida.

ENDANGERED PLANTS

Florida has many endangered ferns. More than 40 of this state's ferns could someday become **extinct.** These include the Florida bristle fern and the royal fern. Another endangered fern is the unusual hand fern. The hand fern can live in only one place: at the ends of dead cabbage palm fronds. Fronds are the long branches and leaves of a palm tree. As more people cut down these dead fronds, the ferns have fewer places to live.

An especially interesting endangered Florida plant is the pitcher plant. The state government believes it is possible that five kinds of pitcher plants may become extinct. These are the white-top pitcher plant, the hooded pitcher plant, the parrot pitcher plant, the decumbant pitcher plant, and the red-flowered pitcher plant.

Pitcher plants are brightly colored and make a sweet nectar that attracts insects. When the bugs land to drink

Pitcher plants sometimes grow up to three feet high. They have long stems with tops that look like open mouths. It looks like this pitcher plant has just received its next meal!

Endangered Plants

Many of the state's endangered living things are plants. If people move into a plant's habitat, the plant can't get up and move somewhere else. Instead, it must find other ways to spread to some other place. It might send its seeds flying into the wind to land in another area. Or it might spread its roots out a long way underground, looking for more water. It is not easy for plants to spread their seeds quickly, and many of them die when people destroy their habitats.

the nectar, they slip and fall into the plant's long stem. Then the plant uses special juices inside it, called **enzymes,** to digest and eat the insect. Pitcher plants need wet areas to grow. Unfortunately, much of their **habitat** has been destroyed by new homes and farms.

ENDANGERED ANIMALS

There are also endangered animals in Florida, including fish such as the shoal bass, the saltmarsh topminnow, and the Okaloosa darter. The Okaloosa darter can only live in seven small streams that begin in Florida's Okaloosa County, which is in the northwest part of the state. These shallow, clear rivers are home to insect **larvae** that the darters like to eat. Humans who build roads and houses in this part of the state could change these streams so much that the Okaloosa darters could no longer live there.

One of Florida's largest endangered animals is the panther. Panthers like to live alone, far from people. Each one needs as much as 275 square miles of space, which is an area about 16.6 miles wide and long. They need this big area to hunt their food, including deer and wild pigs. Today, people have taken over much of the land the panthers once used. There are houses, farms, and roads

Some people who study history think there were more than 1,300 Florida panthers when the first European explorers arrived in 1513. No one is sure exactly how many are left, but scientists believe Florida only has about 50 panthers still living in the wild.

where the panthers used to look for food. Sometimes panthers try to cross these busy roads and are killed by cars.

Other **endangered** animals in Florida include the crocodile, manatee, green sea turtle, and gray bat. As more people move to Florida each year, more animals have trouble finding places to live in the state.

The small Key deer is close to **extinction.** Only about 700 of these deer are still alive, all of them in the Florida Keys. The largest male Key deer are only about 3 feet tall and weigh an average of 80 pounds. Females are smaller. These deer often die after being struck by cars. People sometimes feed the deer along the roads from their car windows. This teaches the deer to think of cars as a source of food. They then wander into the road, where drivers hit them. There are fewer Key deer born each year, as more and more land is used to build homes and other buildings.

SAVING FLORIDA'S WILDLIFE

People in Florida are trying to save the Key deer. The **federal government** set up the Key Deer National Wildlife **Refuge** in 1957. In that year, scientists believed that there were less than 30 Key deer still alive. The government

*The Key deer is the smallest of all white-tailed deer and is not found anywhere else in the world. It is now protected as an endangered **species** in Florida.*

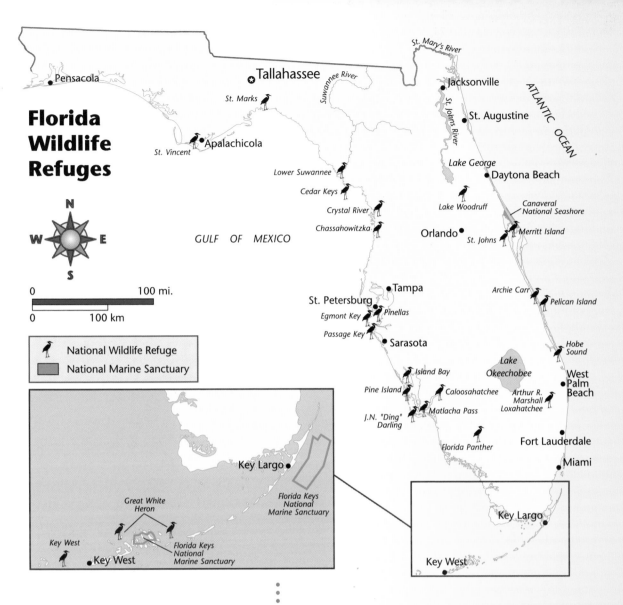

Florida Wildlife Refuges

Legend:
- National Wildlife Refuge
- National Marine Sanctuary

Map labels: Pensacola, Tallahassee, St. Marks, St. Vincent, Apalachicola, Lower Suwannee, Cedar Keys, Crystal River, Chassahowitzka, GULF OF MEXICO, Suwannee River, St. Mary's River, Jacksonville, St. Augustine, ATLANTIC OCEAN, St. Johns River, Lake George, Daytona Beach, Lake Woodruff, Canaveral National Seashore, Orlando, St. Johns, Merritt Island, Tampa, St. Petersburg, Pinellas, Egmont Key, Passage Key, Sarasota, Archie Carr, Pelican Island, Hobe Sound, Lake Okeechobee, West Palm Beach, Island Bay, Pine Island, Caloosahatchee, Arthur R. Marshall Loxahatchee, Matlacha Pass, J.N. "Ding" Darling, Florida Panther, Fort Lauderdale, Miami, Key Largo, Key West

Inset map: Key Largo, Florida Keys National Marine Sanctuary, Great White Heron, Key West, Florida Keys National Marine Sanctuary

now controls more than 8,000 acres of **habitat** in the Florida Keys, where no one can harm the deer.

Wherever you travel in Florida, you are likely to be near a National Wildlife Refuge or a National Marine Sanctuary.

People are also trying to save the Florida panther. One way this is done is by digging tunnels under busy roads. These tunnels let panthers and other animals get where they're going without crossing those roads.

There are other efforts being made throughout the state of Florida to save its valuable wildlife. With those continuing efforts, hopefully humans, animals, and plants can live side-by-side in Florida for many years to come.

Extinct Animals

Many animals that once lived in Florida are now **extinct.** This has happened over thousands of years. We know these animals once existed because scientists have found their **fossils.** Some kinds of animals died because Florida's weather changed about 6,000 years ago, making it impossible for them to find the food they usually ate. The temperatures rose, which made life difficult for animals that were used to the cold. These animals included mastodons and mammoths.

This is a model of a mastodon skeleton that was found in a Florida river. The animal would have had a trunk between its tusks, looking like a hairy elephant.

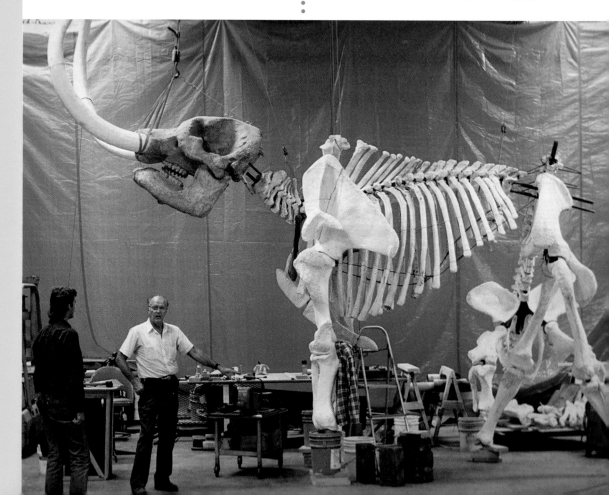

Ancient Armadillos

Florida and other southern states are still home to armadillos. These are slow-moving **mammals** that have short legs and hard shells. They roam around at night looking for dead animals to eat, as well as snakes, ants, and termites.

You may not have known, however, that an enormous type of armadillo lived in Florida 10,000 years ago. It is called the Holmesina, and was as big as a large desk. The Holmesina weighed 600 pounds and was nearly four feet tall.

Then there was the Glyptodont, which was a lot like an armadillo, but was even bigger. The Glyptodont weighed about 2,000 pounds, and was as large as a small car. Both of these animals are extinct today.

Both of these extinct creatures were large and hairy, and they looked a lot like the elephants we know today. They probably died when the weather in ancient Florida became warmer and wetter, and forests grew in places that used to be covered by grassland. The huge animals had used these grasslands for food.

Other extinct Florida animals from long ago include the dangerous saber-toothed cat, which was a large and powerful **carnivore,** as well as camels without humps, rhinoceroses without horns, large vultures with wings fifteen feet long, and huge armadillos.

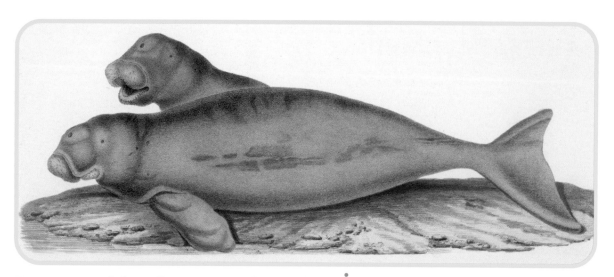

Dugongs used their flippers to push sea grass toward their mouths. They were about 9 feet long and weighed about 600 pounds.

Another **extinct** Florida animal was related to today's **manatee.** It is called the dugong, and it still lives in parts of Asia today. The last Florida dugong died about two and a half million years ago. The dugong looks very

Extinct Sparrow

Not all extinct animals died millions or even hundreds of years ago. One small Florida animal that became extinct very recently was the dusky seaside sparrow. This little bird lived along Florida's central east coast, south of Daytona Beach. However, the United States government built a space center at Cape Canaveral, right in the dusky seaside sparrow's **habitat.** These birds suddenly had no place to live. They could not find food or raise their young, and they began to die.

Walt Disney World Resort in Orlando tried to keep these sparrows alive in captivity. This means that they were kept and cared for by people. Sadly, the birds could not live this way. The last dusky seaside sparrow died at Disney World in June of 1987. Unfortunately, no one will ever see or hear this bird again.

much like the manatee, but the dugong's tail is shaped like a "V" instead of like a fan, as manatee tails are shaped.

This painting shows what scientists think the huge ground sloth looked like. The ground sloth had huge bones, heavy back legs, and a strong tail. This tells us that it could stand on its hind legs to reach high branches and leaves.

Scientists have also discovered the bones of a huge ground sloth that once lived in Florida. A modern sloth is a small, slow-moving **mammal** with long claws. The ground sloths of ancient Florida grew as long as 17 feet, and weighed more than 10,000 pounds. They were larger than today's elephants.

These animals lived in Florida more than two million years ago. They traveled together in groups, called herds, just like elephants. They had very long, sharp claws and long tongues. The ancient sloths ate tree leaves.

People, Plants, and Animals

As we have learned, human beings have changed Florida a great deal. We have taken more and more land from nature. This means that Florida's plants and animals have much more contact with Florida's humans than ever before. Some of these plants and animals are dying. Others have **adapted** to their new life with people.

It is also true that new plants and animals have changed Florida. There are a number of plants and animals that have been brought from other places and that harm Florida's **native species** and even the human population.

*Tropical leafy plants and trees, such as these palm trees, are common around many Florida homes. Palms were in Florida long before humans. In fact, **fossils** of palm leaves have been found that are millions of years old.*

Any person who lives in Florida or visits the state can see signs of increased contact among plants, animals, and humans. Native plants, such as the cabbage palm and saw palmetto, grow around many homes, hotels, and stores all over Florida.

LIZARDS

Some native animals are also very common in Florida homes. Lizards sometimes seem to be all around southern Florida during summer months. People often see them outside on driveways and in gardens, and sometimes even inside their bathrooms and kitchens.

None of Florida's 22 kinds of lizards is poisonous. Some people like to have lizards inside their homes because they eat bugs, including pesky cockroaches. One of these lizards is called the green anole. The male green anole has skin on its throat that it flashes to attract females and scare off other males. This rosy-red skin is called a throat fan, or dewlap. The green anole also changes its skin color from green to brown, so it can blend

Green anoles live in shrubs, grasses, and trees, and are good climbers. Their toes end with a large pad and sharp claw. Each pad has thousands of hairlike bristles that stick to tree bark.

in with trees and not be seen easily by **predators.** Other common Florida lizards that live near people are geckos and skinks. Skinks have shiny scales and thick, round bodies. The southeastern five-lined skink causes cats that eat them to become sick. When cats learn this happens, they stop eating other skinks.

ALLIGATORS

As people continue to build homes farther west in southern Florida, they take over more land that once was a part of the Everglades. This is why alligators sometimes end up in backyard swimming pools at these homes. Alligators also find their way into canals, streams, and ponds near human beings. Sometimes the alligators are looking for food. Sometimes they are only looking for a wet place to go. They don't usually harm people, unless the people are bothering the alligators in some way. However, hungry alligators have been known to eat dogs that got too close to the water.

EXOTIC LIFE

Another way that people have changed nature in Florida is by bringing new plants and animals to the state. These are called "exotics," which means that they are not **native** to Florida. For example, the state has about 6,000 different kinds of plants today. Only 2,500 of these are native plants. That means people brought 3,500

This four-foot-long alligator is sunning itself near an apartment building in Florida.

The Melaleuca Problem

The melaleuca tree, an exotic plant, has created one of Florida's worst problems. Melaleuca trees drink a lot of water. One acre of melaleucas soaks up 2,100 gallons of water each hour. They also spread quickly, and new trees grow easily.

People brought the melaleuca from Australia to Florida in 1906, when they wanted to dry out the Everglades. They hoped to build homes and farms there. These people didn't understand the importance of the Everglades.

Now people want to get rid of Florida's melaleucas, because the trees keep spreading and drinking valuable water in the Everglades and other parts of southern Florida.

kinds of plants to Florida through the years and allowed or helped them to grow here. Some of these exotic plants and animals have caused a lot of trouble.

AUSTRALIAN PINE

Florida's problem plants include the Australian pine tree and the water hyacinth. The Australian pine is seen very often, especially in southern Florida. These trees grow very quickly and very tall, rising high over homes. This pine also spreads quickly.

Australian pines are weak trees. Their branches may die during freezes. Also, entire trees can fall easily when they get very tall and begin to lean in one direction. This means that branches, or even whole trees, can fall on homes or on people. These trees drop so many of their long needles that they cover the ground. This makes it hard for other plants to grow there.

The Spottile Canal in Brevard County, Florida, is overgrown with water hyacinths. Although they look pretty, water hyacinths are a big problem in Florida's inland waters. Diseases and insects control the growth of water hyacinths in South America, where the plants first grew. But in Florida and other regions where people have brought the plant, there is nothing in nature to control its growth.

WATER HYACINTH

Florida's government has spent a lot of money trying to get rid of another problem plant called the water hyacinth. This plant was carried to Florida from South America by a person who liked the way it looked.

The water hyacinth floats in the water and has pretty purple flowers. It grows so fast and gets so thick, however, that boats cannot pass through. The water hyacinth now blocks many of Florida's streams, rivers, and lakes.

EXOTIC ANIMALS

Exotic animals have also caused many problems for Florida. For example, parrots, which many people think are pretty birds, now fly over many cities in southern Florida. They are not **native** to the state. They actually

European Starling

People brought the first starlings into the United States from Europe in 1890. They have become big pests in Florida.

In cold-weather months, starlings travel in very large groups and eat grain planted by farmers. They are very aggressive birds, which means they often chase native birds away from their food and nests. This has forced native birds such as woodpeckers to move from their habitats.

escaped from birdcages and began to spread throughout the area. Now these wild parrots force native birds from their **habitats.**

INSECTS

Insects are among the other exotic animals causing trouble for Florida. The red fire ant is one of the worst problems. These small, red ants were somehow brought from Brazil into the state of Alabama around 1940. No one is sure how this happened, but by 1957 the ants were in Florida. Today these pests are found all over the state.

The ants build large mounds of dirt, where they live by the thousands. These ant mounds can ruin lawns and damage lawn mowers. The biggest problem is that red fire ants bite people and animals.

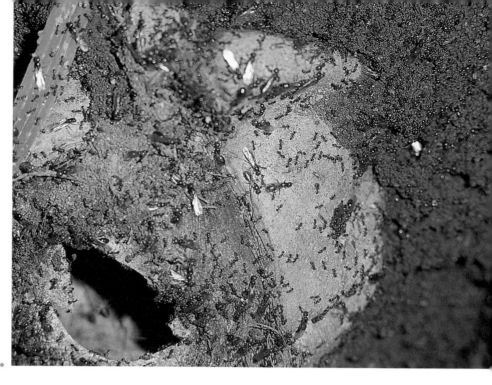

Florida homeowners often apply special chemicals to destroy red fire ant mounds like the one here. Hundreds of thousands of red fire ants can live in one mound.

If a person or animal, such as a cow, steps on a red fire ant mound, many ants immediately race out to bite the intruder. Red fire ant bites usually hurt. Sometimes cows and other animals get so many bites that they die. Even people sometimes die from red fire ant bites.

The Asian tiger mosquito is another exotic pest in Florida. This biting insect came by ship from Taiwan or Japan to Florida in 1986. Today it can spread a disease that makes both horses and people sick. The disease is called eastern equine encephalitis.

RESPECT FOR NATURE

By now, you understand that Florida's **habitats** are easily damaged by human beings. People usually do not know at first when they are doing something to hurt nature so badly, but changes people have brought to the state often make it hard for **native** plants and animals to live here. Each of us can learn something from this. We can try to respect nature, and to help save it for the future.

In Florida and everywhere else, we should always treat plants and animals with great care. If we do, they will always be around for us to enjoy.

Map of Florida

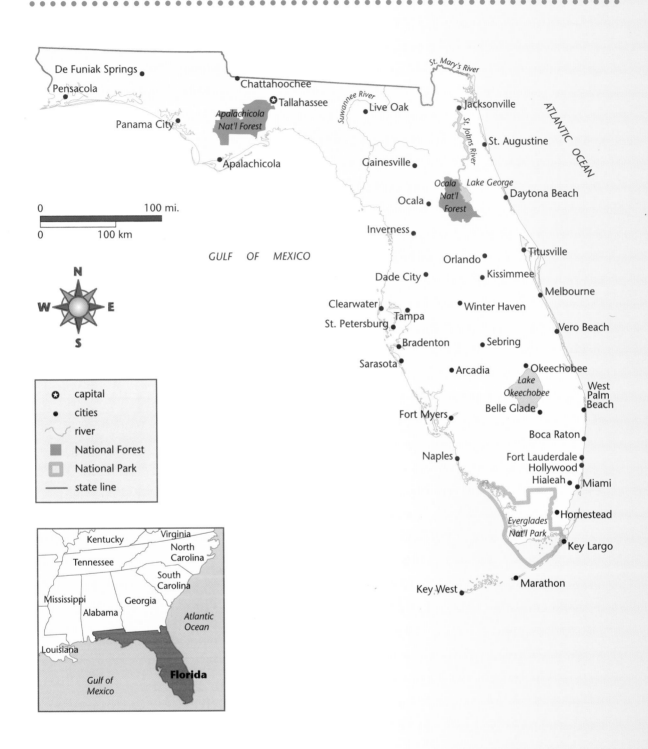

Map labels:

De Funiak Springs
Pensacola
Chattahoochee
Tallahassee
Apalachicola Nat'l Forest
Panama City
Apalachicola
Suwannee River
Live Oak
St. Mary's River
Jacksonville
St. Johns River
St. Augustine
ATLANTIC OCEAN
Gainesville
Ocala Nat'l Forest
Lake George
Daytona Beach
Ocala
Inverness
Titusville
Orlando
Kissimmee
GULF OF MEXICO
Dade City
Melbourne
Clearwater
Winter Haven
Tampa
St. Petersburg
Vero Beach
Bradenton
Sebring
Sarasota
Arcadia
Okeechobee
Lake Okeechobee
West Palm Beach
Fort Myers
Belle Glade
Boca Raton
Naples
Fort Lauderdale
Hollywood
Hialeah
Miami
Everglades Nat'l Park
Homestead
Key Largo
Key West
Marathon

Scale:
0 — 100 mi.
0 — 100 km

Compass:
N W E S

Legend:
- ✪ capital
- • cities
- ⌇ river
- ▪ National Forest
- ▫ National Park
- — state line

Inset map:
Kentucky
Virginia
Tennessee
North Carolina
South Carolina
Mississippi
Alabama
Georgia
Atlantic Ocean
Louisiana
Florida
Gulf of Mexico

Glossary

adapt to change in a way that allows an animal or plant to live in new conditions. An adaptation is such a change.

amphibian animal, such as a turtle or a frog, that lives in and around both land and water

carnivore animal that eats other animals in order to live

consumer living creature that needs to eat other living things to survive

continental United States any of the 48 states that all touch other states; all of the United States except for Alaska and Hawaii

coral polyp tiny sea creature that lives in large groups of other polyps to form coral reefs

ecosystem all of the animals and plants that make up a particular community living in a certain environment

endangered put at risk or in danger; in danger of dying out

environment all the things that surround a person, animal, or plant and affect those living things

enzyme liquid in a plant or animal that helps digest food

epiphyte plant that grows on or is attached to another plant because it needs to be supported; also called an air plant, it is not attached to the ground in any way

evolve to change gradually, over time, because of different forces in the environment

extinct no longer existing

federal government elected representatives of the people of the United States who work in Washington, D.C., and make decisions regarding all states in this country

food chain diagram of the plants and animals that need each other for food within a particular habitat

fossil remains or traces of a living thing of long ago

fragile easily broken or ruined

gill breathing organ found on most water-dwelling animals, including fish

habitat place that is just right for a certain group of animals and plants to live

herbivore animal that eats only plants in order to live

larva young form of an insect, which looks like a worm and has no wings. Larvae is the plural form of larva.

mammal warm-blooded animal that has a backbone. Female mammals produce milk for feeding their young. Dogs, whales, mice, and human beings are examples of mammals.

manatee large, slow-moving mammal that lives in warm waters and eats only plants

microorganism living thing that is so small that it can only be seen with a microscope

native plant or animal that is originally from an area

natural resource anything found in nature that humans consider necessary, like air, plants, animals, minerals, and water. Natural resources can be renewable (they can make more of themselves, like plants, if given the chance) or nonrenewable (they cannot make more of themselves, like coal)

predator living creature that hunts and eats other living beings in order to survive

producer living thing that makes its own food

reef underwater crest of rocks, coral, or sand that is near the surface of the ocean

refuge area of land where animals and plants are protected

reptile animal, such as a lizard or snake, that is cold-blooded and sheds its skin regularly

shellfish oysters, clams, and other small, soft sea animals that live in hard shells

species group of plants or animals that are alike in certain ways

tadpole young animal that will grow into a frog

tentacle on a coral polyp, tiny arm-like limbs that catch food and pull it toward the polyp's mouth

turpentine particular oil from a tree that is used as paint thinner

unique unlike, or different from, anything else

More Books to Read

Clark, Margaret G. *Save the Florida Key Deer.* N.Y.: Penguin Putnam Books for Young Readers, 1998.

Clark, Margaret G. *The Threatened Florida Black Bear.* N.Y.: Penguin Putnam Books for Young Readers, 1995.

Lantz, Peggy S., and Wendy A. Hale. *The Florida Water Story: From Raindrops to the Sea.* Sarasota, Fla.: Pineapple Press, Inc., 1998.

An older reader can help you with this book:

Lantz, Peggy S. and Wendy A. Hale. *The Young Naturalist's Guide to Florida.* Sarasota, Fla.: Pineapple Press, 1994.

Index

About the Author

Bob Knotts is an author and playwright who lives near Fort Lauderdale, Florida. He has published 24 novels and nonfiction books for both young readers and adults. He also writes for several top national magazines.